TECHNICAL REPORT

The Value and Impacts of Alternative Fuel Distribution Concepts

Assessing the Army's Future Needs for Temporary Fuel Pipelines

David M. Oaks • Matthew Stafford • Bradley Wilson

Prepared for the United States Army
Approved for public release; distribution unlimited

RAND ARROYO CENTER

The research described in this report was sponsored by the United States Army under Contract No. W74V8H-06-C-0002.

Library of Congress Cataloging-in-Publication Data is available for this publication.

ISBN 978-0-8330-4666-6

The RAND Corporation is a nonprofit research organization providing objective analysis and effective solutions that address the challenges facing the public and private sectors around the world. RAND's publications do not necessarily reflect the opinions of its research clients and sponsors.

RAND® is a registered trademark.

Cover photo courtesy of the 240th Quartermaster Battalion, U.S. Army.

Published 2009 by the RAND Corporation
1776 Main Street, P.O. Box 2138, Santa Monica, CA 90407-2138
1200 South Hayes Street, Arlington, VA 22202-5050
4570 Fifth Avenue, Suite 600, Pittsburgh, PA 15213-2665
RAND URL: http://www.rand.org/
To order RAND documents or to obtain additional information, contact
Distribution Services: Telephone: (310) 451-7002;
Fax: (310) 451-6915; Email: order@rand.org

Preface

After the end of the Vietnam War, the Army developed an improved capability to emplace above-ground, temporary petroleum pipelines for providing wholesale fuel support to all U.S. land-based forces, including Air Force, Marine Corps, and Navy forces ashore. Yet this petroleum pipeline capability was put into operation in only one of the two major combat operations in the past 30 years. There is some question as to whether this single employment was feasible only because of a unique set of circumstances unlikely to be present in future situations in light of the expected expeditionary nature of anticipated contingencies. The examination of this question was the focus of the project entitled "The Value and Impacts of Alternative Fuel Distribution Concepts."

The purpose of this report is to document project findings that inform the U.S. Army on the anticipated future requirements for a petroleum pipeline capability, provide an assessment of a range of options for meeting those requirements, and offer recommendations contingent on the decision maker's appraisal of future conditions. These findings should be of interest to those engaged with future Army logistics support force structure requirements.

This research was sponsored by Lieutenant General John M. Curran, the Deputy Commanding General, Futures/Director, Army Capabilities Integration Center of the United States Army Training and Doctrine Command, with oversight provided by Major General Mitchell H. Stevenson, Commanding General, United States Army Combined Arms Support Command. It was conducted within RAND Arroyo Center's Military Logistics Program. RAND Arroyo Center, part of the RAND Corporation, is the Army's federally funded research and development center for policy studies and analyses.

The Project Unique Identification Code (PUIC) for the project that produced this document is ATFCR07226.

For more information on RAND Arroyo Center, contact the Director of Operations (telephone 310-393-0411, extension 6419; FAX 310-451-6952; email Marcy_Agmon@rand. org), or visit Arroyo's web site at http://www.rand.org/ard/.

Contents

Figures

Tables

Summary

The Army maintains the capability to employ temporary petroleum pipelines. With the fiscal year (FY) 08–13 program objective memorandum (POM) force, the Army proposes to retain two Active and twelve Reserve Petroleum Pipeline and Terminal Operating (PPTO) companies. There is the potential to convert up to four of the PPTO companies to a unit design centered on the in-development Rapidly Installed Fuel Transfer System (RIFTS) technology, with the remaining companies retaining the existing Inland Petroleum Distribution System (IPDS) system.

But this temporary pipeline capability has been put into operation in just one of the two major combat operations of the past 30 years. Moreover, as the Army transforms, the question arises whether this single pipeline deployment was feasible only because of a unique set of circumstances unlikely to recur in anticipated future expeditionary and nonlinear warfare. If so, there is a further question of whether the Army should reallocate the resources associated with the pipeline force structure to fill other force structure needs. This report attempts to answer these questions, starting with a review of historical and anticipated requirements for temporary pipelines and then moving on to an assessment of existing and future unit designs to meet future requirements.

The review of historical pipeline use since Vietnam is summarized in Table S.1. Looking at these instances together as a group, an interesting pattern emerged. Pipelines tended to be used at discrete groups of distances:

- Short cases of 25 miles or less;
- Middle distances of about 50 miles; and
- Long distances of over 100 miles.

Further, the short-distance uses were the majority, with a few middle-distance occurrences, and just a single long-distance employment.[1]

A review of future pipeline requirements incorporated into war plans and security planning scenarios revealed a pattern very similar to the historical one. As shown in Table S.2, there are several short-distance requirements from less than 3 up to 25 miles, some mid-distance requirements around 50 miles, and again a single long-distance requirement of 160 miles.

There are several fluid transfer systems with the potential to meet the anticipated future requirements. As already mentioned, the existing (legacy) system is the IPDS. It is primarily comprised of 19-foot-long aluminum pipe sections that are 6 inches in diameter. It is high pressure, 740 pounds per square inch (psi), and capable of delivering up to 1 million gallons per

---

[1] The other long-distance case in this list was only partially complete at the end of hostilities and therefore not counted.

Table S.1
Summary of Historical Cases from Vietnam to OIF

	Short Distance Up to 25 Miles	Medium Distance ~50 Miles	Long Distance 100 Miles +
Vietnam	DONG NAI–LONG BIN AB • 4 miles: 6″ LWST (3 parallel) POL PIER–CAM RANH BAY AFB • 6 miles: 6″ LWST (2 parallel) SAIGON–TAN SAN NHUT AFB • 6 miles: 6″ LWST (2 parallel) PHAN RANG–PHAN RANG AFB • 10 miles: 6″ LWST (2 parallel) QUI NHON–PHU CAT AFB • 17 miles: 6″ LWST VUNG RO BAY–TUY HOA • 18 miles: 6″ LWST (2 parallel)	QUI NHON–AN KHE • 50 miles: 6″ LWST AN KHE–PLEIKU • 59 miles: 6″ LWST	
Desert Shield	RAS TANURA–KING FAHD APT • 25 miles: 6″ IPDS (contaminated, not used)		ADDAMMAM-HAFIR AL BATIN • 260 miles: 6″ IPDS (partially complete at time of cease fire)
Somalia	MOGADISHU: PORT–AIRFIELD • 2.5 miles: 6″ IPDS		
Iraqi Freedom		USMC: BP WEST–LSA VIPER • 54 miles: 6″ HRS	UDARI–TALLIL • 160 miles: 6″ IPDS
Total	7 (8)	3	1 (2)

NOTES: HRS = Hose Reel System, AB = air base, AFB = Air Force Base, LWST= lightweight steel tubing, IPDS = Inland Petroleum Distribution System.

day. A PPTO company is the Army unit designed to operate up to 90 miles of IPDS pipeline.[2] Emplacement of 90 miles of pipeline would take about a month using the planning factor of 2 to 3 miles per day. The IPDS system also contains organic storage capacity up to 3.78 million gallons in fabric bags.

The developmental replacement for IPDS is the RIFTS. Two key differences from IPDS are that (1) the RIFTS uses a flexible 6-inch hoseline that is expected to be as capable as IPDS in terms of throughput; and (2) the emplacement rate is projected at 20 miles per day. In testing to date, the RIFTS hose has not yet achieved high-enough pressure to make it as capable as IPDS, and its unit design does away with the organic storage capability.

Similar to the RIFTS concept is the Marine Corps' Hose Reel System (HRS). It, too, is based on a flexible 6-inch hoseline that can be rapidly emplaced. HRS operates at low pressure but is a proven system already used in combat. It has the organic capability to store over 1.1 million gallons. A fourth delivery capability, a 7,500-gallon tanker truck company, was included in the comparative analysis, as this is an alternative to employing temporary pipelines.

---

[2] An Engineer Pipeline-Construction Support Company is the unit doctrinally tasked to emplace temporary pipelines, as described in Headquarters, Department of the Army, Field Manual 5-482, *Military Petroleum Pipeline Systems*, Washington, D.C.: Department of the Army, August 26, 1994, p. 1-8.

Table S.2
Future Scenarios

Requirement	Length (miles)	Event	Unclassified Description	Time Sensitive?
A	25 (likely requiring multiple lines)	ISB	Permissive but remote location, time sensitive, very high throughput required	Yes
B	50	MCO	Time-sensitive requirement to move POL	Yes
C	50	MCO	Time-sensitive requirement to move POL	Yes
D	10	NEO	Austere environment, potential requirement to support other nations' forces as well	Yes
E	10	HA/HLD	Austere environment, respond to an environmental disaster	Yes
F	160	MCO	Long distance pipeline, not time sensitive	No
G	35 (likely requiring multiple lines)	MCO	Not time sensitive, very high throughput	No
H	10	MCO	Short distance, not time sensitive	No

These four systems, IPDS, RIFTS, HRS, and tanker trucks, were assessed in their performance against four key future scenario types: intermediate staging base (ISB), noncombatant evacuation operation (NEO), time-sensitive major combat operation (MCO), and non-time-sensitive MCO. The evaluation metrics are strategic deployability, number of soldiers required, time to emplace,[3] and potential investment cost. This analysis yields no clear winner among the four systems. IPDS does well in cost and soldiers required but is slow to emplace. Trucks are deemed an infeasible solution for two of the scenarios and are much less efficient as the fuel delivery distance increases. RIFTS is the fastest to emplace but the most expensive and with technical performance development hurdles still to be overcome. Finally, the HRS is economical and fast to emplace but becomes, like trucks, more inefficient as delivery distance increases in that an increasing number of pump stations and associated personnel are required.

Therefore, the overall recommendation is to proceed within the context of the decision maker's most important concern, as shown in Table S.3. If cost, for example, is the most pressing issue, then the best choice is to retain the IPDS system, perhaps supplemented by some limited acquisition of HRS hoseline. Alternatively, if strategic mobility of pipeline assets is the most important issue, then acquisition of RIFTS or the selective prepositioning of IPDS assets are the best options.

Temporary pipelines remain an attractive capability to retain in the force structure, but the question is how much. Unlike many support requirements, though, temporary pipelines do not appear to have a rotational requirement. It is generally not cost-effective to employ a pipeline unit if the fuel requirement is small or infrequently required. And as they pose an

---

[3] The use of "time to emplace" as a key measure may be seen as less compelling than other measures, such as "gallons delivered per unit time." In the context of this research, though, the alternative newer system the Army was seriously considering, RIFTS, was promoted primarily due to its promised speed of emplacement while holding delivery volume essentially constant. This was to be the overriding "selling point" and most-often-cited parameter in various war plans (which often do not make reference to a required number of gallons of fuel to be delivered over time).

Table S.3
Choice in Light of Most Pressing Concern

If highest concern is . . .	. . . then choose
Cost	IPDS (maybe HRS)
Mobility	RIFTS (or prepo IPDS)
Time to employ	RIFTS
Personnel	IPDS + RIFTS
Technical risk	IPDS + trucks

obstacle to maneuver of both military and indigenous traffic and, in contested areas, present an inviting target for enemy mischief or theft and pilferage, it is not desirable to keep them in place too long. And by their temporary nature in design and materials, they degrade over time, leaking or failing.

So one approach to take to estimate the total amount of temporary pipeline equipment and units needed is to assess how much might be needed simultaneously based upon which future scenarios might occur simultaneously. Figure S.1 indicates that about 500 miles of pipeline capability could cover all requirements at the same time, providing one estimate. Naturally, a rational case can be made for lesser totals or for dividing totals by capability. For example, if the Army decides to continue acquisition of a hoseline-based system, a reasonable amount could be 220 miles, an amount to cover the more time-sensitive scenarios, leaving the legacy IPDS systems to cover the remaining 275 miles of less time-sensitive contingency requirements.

The Way Ahead for the Army

Temporary pipeline capabilities do not come without costs. Pipelines take time and resources beyond the petroleum pipeline units, notably engineer support, to set up and operate. Allocating acquisition dollars to new technology, primarily flexible hoseline, and improved pumping stations can lessen emplacement time and the engineering support required. The Army should consider focused investments in these areas. Similarly, the need to protect the pipeline against pilferage or sabotage also remains, a task the Military Police are doctrinally expected to perform but one typically beyond their ability to cover due to other demands in theater.[4] Reorganization of the personnel allotted to the existing PPTO Company can make that unit more capable of self-protection, again, a step the Army should consider taking, as is the possibility of merging PPTO, Assault Hoseline, and Tactical Water Distribution System teams into one fluid-transfer-capable unit type (that is, a multi-function unit capable of operating either POL or water equipment but not defined by its equipment type).

---

[4] While the numbers of Military Police (MP) units that could potentially be needed to support pipeline operations may be of interest, this study does not address this question for the following reasons. Of the scenarios surveyed, several are in permissive environments, which indicates that MP support is not always necessary. Further, the rules of allocation for MP units in the current Total Army Analysis process include no direct link between units, such as the Military Police Combat Support Company, and a requirement to protect a distance of pipeline.

Figure S.1
Sizing Total Pipeline Requirement by Simultaneity

Requirement	Length (miles)	Event
A	75 (3 × 25)	ISB
B	50	MCO
C	50	MCO
D	10	NEO
E	10	HA/HLD
F	160	MCO
G	105 (3 × 35)	MCO
H	10	MCO

195 miles (more time sensitive)

275 miles (less time sensitive)

~195 + training set = 220-mile RIFTS buy

~700 miles of IPDS today

470 miles (if all occurred simultaneously)

RAND TR652-S.1

Acknowledgments

The authors would like to thank Colonel Shawn Walsh for sharing his experience as commander of the unit that operated the IPDS pipeline during OIF-1 and for other advice and insights on pipeline operations. At CASCOM, Tim Trauger, Mari Wells, Major Don Herko, William Perdue, Charles Burden, and Colonel Dan Mitchell were very helpful in providing historical data, technical information, and feedback during in-progress reviews. Mike Pendergast from the office of the Army Deputy Chief of Staff, G4, provided very helpful access to and suggestions on the use of future planning scenarios.

Special thanks go to several at the Radian Corporation and the Marine Corps. At Radian, Kevin Stump and Ed Martin were very generous in sharing their personal experience and technical analysis of various pipeline systems, and in making time available both at their offices and in the field at the FORSCOM pipeline training area at Camp Pickett, Virginia. Joe Irwin of the Marine Corps Systems Command was likewise very willing to take time to explain the organization, equipment, and experience of the Marine Hose Reel System used in Iraq. Other Marines helped tell this story, most notably Mike Giambruno.

At RAND, Jerry Sollinger provided assistance with organizing this document, and Dave Orletsky, Eric Peltz, and Rick Eden provided thorough reviews. Errors of fact or interpretation remain the responsibility of the authors.

List of Acronyms and Abbreviations

CASCOM	Combined Arms Support Command
COCOM	Combatant Command
FORSCOM	Forces Command
gpm	gallons per minute
HA	Humanitarian Assistance
HEMTT	Heavy Expanded Mobility Tactical Truck
HRS	Hose Reel System
IPDS	Inland Petroleum Distribution System
ISB	Intermediate Staging Base
ISO	International Organization for Standardization
LSA	Logistics Support Area
LWST	Lightweight Steel Tubing
MAOP	Maximum Allowable Operating Pressure
MCO	Major Combat Operation
MP	Military Police
NEO	Noncombatant Evacuation Operation
ODS	Operation Desert Storm
OIF	Operation Iraqi Freedom
OPDS	Offshore Petroleum Distribution System
POL	Petroleum, Oil, and Lubricants
PPTO	Petroleum Pipeline and Terminal Operating (Company)
RC	Reserve Component
RIFTS	Rapidly Installed Fluid Transfer System
RSOI	Reception, Staging, Onward movement, and Integration
TAA	Total Army Analysis
TAFDS	Tactical Airfield Fuel Dispensing System
TACOM	Tank-automotive and Armaments Command
TO&E	Table of Organization and Equipment
TPT	Tactical Petroleum Terminal
TWDS	Tactical Water Distribution System

CHAPTER ONE

Introduction

Background

Pipelines are the most efficient means for moving large volumes of liquid products. In the United States they move two-thirds of the oil transported annually, roughly thirteen billion barrels of both crude and refined petroleum products, safely and at a cost much lower than railroads, trucks, or barges can offer.[1] Because of these positive attributes, today the United States has the densest network of petroleum pipelines in the world.

Like the commercial sector, the U.S. Army has had a history of investing in the personnel and materiel that enable it to construct and operate petroleum pipelines. With a general planning estimate for half of all sustainment tonnage moved in a theater of operations to consist of petroleum products, military use of pipelines is an attractive proposition.[2] Moreover, Joint doctrine for bulk petroleum clearly names pipeline distribution as the preferred method for inland petroleum distribution.[3]

The Army's pipelines differ from those used commercially, however, in that they are above-ground systems purposely designed for deployment, easy emplacement, operation, and retrieval. The currently fielded pipeline system dates to the period just after the Vietnam War, when the Army found itself with only 25 miles of non-mission-capable petroleum distribution equipment on hand.[4] To resolve this problem, an improved pipeline set with then state-of-the-art technology was developed in the mid-1970s, together with an associated unit to operate it.

This improved pipeline, called the Inland Petroleum Distribution System (IPDS), can be much more quickly emplaced than the World War II vintage lightweight steel tubing (LWST) system it replaced, and can deliver larger amounts of petroleum.[5] Among the many technical improvements to the IPDS were that each pipe section is aluminum and weighs 110 pounds, making it man-portable, and assembly was made much easier in that pipe sections are joined with a simple hammer-driven pin as opposed to the two bolts and nuts required for the LWST

---

[1] Association of Oil Pipelines, Safety Record. As of January 2009, navigable from: http://www.aopl.org/go/site/888/

[2] Headquarters, Department of the Army, Field Manual 10-67, *Petroleum Supply in Theaters of Operations*, Washington, D.C., February 18, 1983, p. 2-1.

[3] Joint Chiefs of Staff, Joint Publication 4-03, *Joint Bulk Petroleum and Water Doctrine*, Washington, D.C.: Joint Chiefs of Staff, May 23, 2003, p. I-4.

[4] Kevin Born, "Short History of Tactical U.S. Military Pipelines," *An Assessment of the Rapidly Installed Fluid Transfer System (RIFTS)*, Alexandria, VA: Radian, Inc., January 2004, Appendix A.

[5] Keith E. Mattox, "The Army's Inland Petroleum Distribution System," *Quartermaster Professional Bulletin*, Spring 1998.

system. Strategic mobility was improved, as each IPDS pipe is just 19 feet in length, making it able to fit inside a 20-foot International Organization for Standardization (ISO) container, and fewer pump stations were needed because the IPDS has a higher maximum allowable operating pressure (MAOP) of 740 pounds per square inch (psi), as compared to the 600 psi of the LWST system.[6] Today, the Army has roughly 700 miles of IPDS conduit (in various states of readiness) and 14 company-sized pipeline operating units (in both the Active and Reserve components) in its inventory.

Motivation for the Study

Since its creation roughly 30 years ago, the IPDS pipeline system has been used at its full design potential only once, during Operation Iraqi Freedom. There is a concern that this singular use might have been an artifact of the particularly favorable set of conditions. The nature of the specific theater, which has relatively open and flat terrain and nearby large petroleum refining facilities, facilitated the use of IPDS. Additionally, certain aspects of the operation—including a friendly host nation, long preparation time, and the ability to safely preposition assets and personnel—were ideal for the use of IPDS. These conditions are different from the expectations driving a significant amount of Department of Defense planning, with expectations of the need to handle nonlinear battlefields, expeditionary operations, and irregular forces.

Over the past few years, force developers at the Combined Arms Support Command (CASCOM) and materiel developers at the Tank-Automotive and Armaments Command (TACOM) have created a new unit design and associated prototype equipment for a pipeline system that promises improvement over the IPDS. The key difference is the replacement of the IPDS's rigid pipe sections with a flexible hoseline carried in a vehicle-mounted motorized drum. It is hoped that this system, known as the Rapidly Installed Fluid Transfer System (RIFTS), will reduce long-distance pipeline emplacement times from days to hours, and likewise will be able to be much more quickly retrieved and emplaced again elsewhere on the battlefield. It is this attribute of speed of emplacement that designers see as the key response to keep petroleum pipelines viable in a future expeditionary, nonlinear combat environment. The big drawbacks of the RIFTS are that its prototypes have not yet met anticipated performance in terms of MAOP within the hoseline and its high cost, due to both the advanced technology of the conduit itself and the dozens of large vehicles (Heavy Expanded Mobility Tactical Truck [HEMTT]-sized) in the unit design. These concerns and higher priority funding needs in the Army budget have slowed materiel development of the RIFTS system and slowed support for migration to the RIFTS unit design.

The obvious alternative to temporary pipelines is truck units, either ones organic to the Army, furnished by coalition partners, or those contracted for with commercial providers. The Army is already taking some risk with its pool of organic truck units in that it has fewer units than its planning process indicates it may require to meet future wartime demands.[7] Counting

---

[6]   Born, Appendix A.

[7]   For example, the Total Army Analysis (TAA) 08–13 results indicated a requirement for 55 Medium Truck Companies, Cargo (echelon above corps or EAC), but only 50 were resourced in the POM 08–13 Army Structure Message in April 2006.

on support from partners or civilian providers is not a preferred alternative early in a conflict, because quick access to such assets cannot be assured.

Thus, the Army faces tough choices on how to distribute its allotted share of manpower and materiel investment resources to units. Should it continue with its current set of IPDS equipment and units and perhaps miss an opportunity to reinvest some or all of the resources they represent against truck unit shortages? Alternatively, should the Army take today's pipeline resources and reinvest part or all in the new but unproven RIFTS design? Or are there other alternatives?

How This Report Is Organized

The report is laid out as follows. Chapter Two reviews the historical use of pipelines from the Vietnam War up through and including Operation Iraqi Freedom, and examines potential future fuel pipeline requirements based upon modeling scenarios, existing combatant command (COCOM) operation plans, and illustrative planning scenarios. From this review of history and anticipated needs, a picture of expected future demands for pipelines emerges. Chapter Three provides an assessment of existing and future units and technologies to meet these demands, including an estimate of some relative costs to pursue each of these technologies/unit types. Finally, the report concludes with findings and policy recommendations.

CHAPTER TWO

Pipeline History and Anticipated Requirements

This chapter begins with a concise history of the use of temporary pipelines from the Vietnam War through Operation Iraqi Freedom. This look at the occurrences of pipeline use in a variety of circumstances over the past 30 years provides insights about the potential future use of deployable pipelines. It is reasonable to assume that pipelines may again be employed under circumstances similar to those in which they have been employed in the past, so it is important to know what these circumstances are. The discussion then turns to planning for future operations. From a review of combatant command contingency plans, Office of the Secretary of Defense security posture scenarios, and combat models used in the most recently completed force structure analysis (the Total Army Analysis, or TAA), we assemble a broad view of potential future pipeline requirements. The likely scenarios combined with past pipeline uses may provide insights about the value of pipelines in future operations.

Vietnam

In the mid-1960s, as U.S. military involvement in Vietnam grew, the Army employed various methods of fuel delivery. At the start, small military tanker trucks and local commercial vehicles of up to 3,000-gallon capacity were used. This means of delivery was soon overwhelmed by growing demand. Eventually, petroleum distribution in Vietnam included tanker ships, barges, and even aircraft (mainly carrying 55-gallon fuel drums) in addition to tanker trucks.[1] Additionally, to meet increased demand while reducing reliance on truck and aircraft delivery, the Army chose to install pipelines wherever there was large steady demand and where the lines could reasonably be secured.[2]

As shown in Figure 2.1, most of the pipelines in Vietnam covered short distances, not exceeding 18 miles. A total of 233 miles of pipe was installed, much of it World War II vintage 6-inch lightweight steel tubing (LWST) with bolted couplings. The majority of these pipelines spanned short distances to serve airfields, such as Long Binh or Tan Son Nhut, and they were often constructed in two or three parallel lines to provide the volume required to sustain a high tempo of aviation operations.

---

[1] Army Quartermaster Foundation, Inc., "The POL Story: To Keep 'Em Running," *Magazine of the 1st Logistical Command*, Vietnam, April 1968.

[2] Joseph M. Heiser, Jr., *Vietnam Studies: Logistic Support*, Washington, D.C.: U.S. Government Printing Office, 1991, pp. 77–78.

Figure 2.1
Petroleum Pipelines in Vietnam

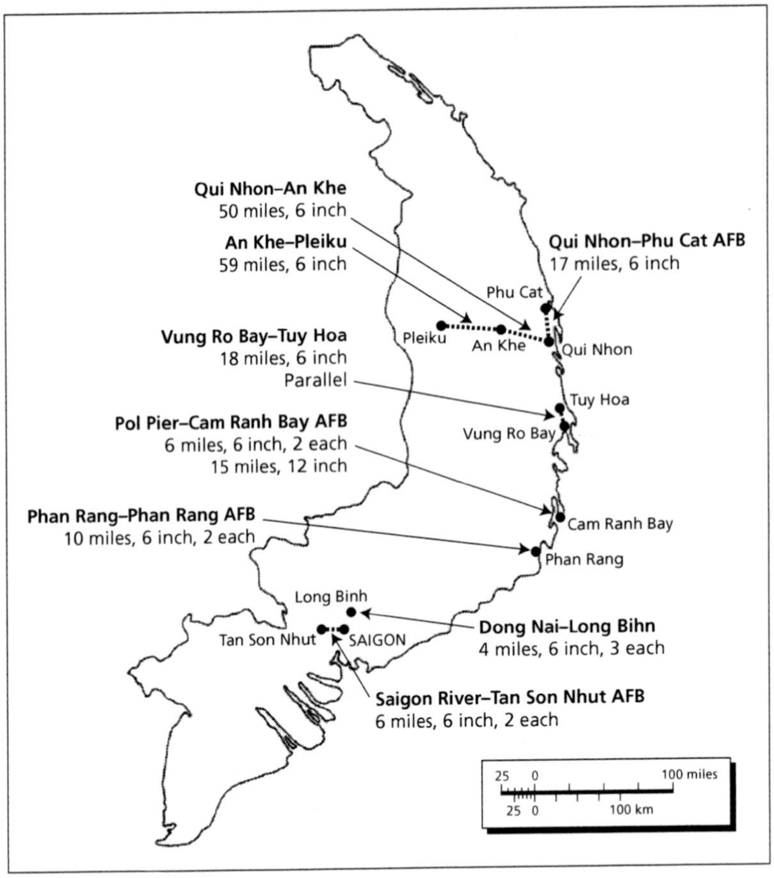

Qui Nhon–An Khe
50 miles, 6 inch

An Khe–Pleiku
59 miles, 6 inch

Qui Nhon–Phu Cat AFB
17 miles, 6 inch

Phu Cat

Pleiku   An Khe   Qui Nhon

Vung Ro Bay–Tuy Hoa
18 miles, 6 inch
Parallel

Tuy Hoa

Pol Pier–Cam Ranh Bay AFB
6 miles, 6 inch, 2 each
15 miles, 12 inch

Vung Ro Bay

Phan Rang–Phan Rang AFB
10 miles, 6 inch, 2 each

Cam Ranh Bay

Phan Rang

Long Binh

Tan Son Nhut   SAIGON

Dong Nai–Long Bihn
4 miles, 6 inch, 3 each

Saigon River–Tan Son Nhut AFB
6 miles, 6 inch, 2 each

25   0   100 miles
25   0   100 km

SOURCE: Lieutenant General Joseph M. Heiser, Jr., *Vietnam Studies: Logistic Support*, Washington, D.C.: U.S. Government Printing Office, 1991, Map 2.
RAND *TR652-2.1*

In the Vietnam record, though, there is one instance of a long stretch of pipeline. This was in fact two segments of roughly fifty miles in length, the first from Qui Nhon to An Khe and the second from An Khe on to Pleiku. The long stretches of forest and countryside traversed by this conduit, together with fact that the LWST was merely bolted together every 20 feet, made it an easy target for theft and sabotage. Fuel losses along this trace were roughly 2.5 million gallons (or about 20 percent of the overall flow) per month.[3]

The short-distance pipelines were more successfully kept secure, likely due to ease of patrolling and maintaining observation along their much shorter legs and due to the lower relative incidence of enemy activity in areas of Vietnam adjacent to major airfields. Thus, an observation that the Army brought out of the Vietnam experience was that pipelines were only efficient if they could be protected.

---

[3] Heiser, pp. 77–78.

Operation Desert Shield/Desert Storm

The end of the Vietnam War found the Army with 25 miles of 1950s vintage petroleum pipeline equipment, none of it mission capable. This circumstance and emerging new requirements coming out of planning for operations in Southwest Asia led to the direction from the Joint Chiefs of Staff to the Army and Navy to develop a more modern deployable bulk fuel distribution capability. The Navy's contribution became the Offshore Petroleum Distribution System (OPDS) designed to transport fuel from a tanker ship to the high water line on the beach. The Army system, the Inland Petroleum Distribution System (IPDS), was developed between 1983 and 1990 to transport the fuel inland from the beach.[4]

Operation Desert Shield (ODS) presented the first opportunity to use the IPDS. Plans were made to construct a pipeline from the Ras Tanura and Al Jubail refineries in the coastal city of Ad Dammam, Saudi Arabia, inland approximately 260 miles to the vicinity of Hafar Al Batin on the Saudi border with Iraq. This pipeline was to feed a series of tactical petroleum terminals (TPTs) sited in Army logistical bases along its trace. A series of events, however, worked against this plan. First, the deployment of the 475th Quartermaster Group, whose personnel were to operate the IPDS, was delayed to allow combat forces to get to the theater first. Similarly, major portions of the IPDS equipment were delayed in shipment to Saudi Arabia, again due to the priority for cargo movement going to combat units. Finally, the Saudi government was slow to approve the proposed right-of-way for the above-ground pipeline. This final point was not a trivial one in that the IPDS, as an above-ground system, did pose an obstacle to movement and a potential environmental hazard, even in a mostly barren desert region.[5]

Despite these delaying events, two sets of IPDS pipeline were emplaced. The first spanned the 25 miles between the Ras Tanura refinery and King Fahd airport. An attempt was made to put this section into operation; however, the initial batch of fuel pumped into it was contaminated, after which this pipeline was shut down. The second IPDS pipeline consisted of two runs, the first about 100 miles from Ad Dammam to Logistics Base Bastogne, followed by another 75 miles continuing on toward Logistics Base Alpha. These sections were never brought into operation, as all work on them was halted with the declared end of offensive operations on February 28, 1991.

Operation Restore Hope: Somalia

The scale of the mission of Operation Restore Hope (ORH) in Somalia was much smaller than ODS, with the construction of just 2.5 miles of IPDS to serve the airfield in Mogadishu. Beside this modest distance, though, there were some interesting aspects of petroleum support there. First, the Navy's OPDS system was used to bring fuel from the tanker, SS *Osprey*, anchored about a half-mile offshore of Mogadishu. Next, it evolved into a joint effort ashore to provide petroleum support in that two Marine Corps systems, the Amphibious Assault Fuel System (AAFS) and the Tactical Airfield Fuel Dispensing System (TAFDS), were also employed to do

---

[4]   Kevin Born, "Short History of Tactical U.S. Military Pipelines," 2004, p. A-5.

[5]   Radian, Inc., *An Assessment of the Rapidly Installed Fluid Transfer System (RIFTS)*, Alexandria, VA: Radian, Inc., January 2004, Appendix A, and Joseph T. Thomas, *Petroleum Operations in the Gulf War: An Operation Desert Storm Personal Experience Monograph*, Strategy Research Project, Carlisle Barracks, PA: U.S. Army War College, April 15, 1993.

retail fuel delivery. Finally, a key consideration underlying the decision to emplace a wholesale pipeline, even over such a distance, was the desire to reduce the exposure of soldiers who otherwise would have been driving fuel trucks at regular intervals through a part of Mogadishu. In 90 days, 4.7 million gallons of fuel were delivered through this combination of systems.[6]

Operation Iraqi Freedom

There were two significant employments of temporary pipelines in Operation Iraqi Freedom (OIF). First was the emplacement of roughly 160 miles of IPDS in Iraq, the singular use of this system in a major contingency. The second, perhaps less well known, is the Marine Corps' deployment of its Hose Reel System (HRS).

Marine Corps Hose Reel System in OIF

In order to cover the time it would take the Army to emplace and bring into operation its IPDS system during the opening phase of OIF, the Marine Corps decided to temporarily employ its retail Hose Reel System in a wholesale fashion. Marine fuel support planning envisioned the emplacement of the HRS from the Kuwaiti border northwest to a Forward Operating Base at Jalibah, Iraq, to free up fuel trucks to concentrate forward on the fight toward Baghdad. While Marine Corps engineering units had experimented with HRS in Southern California prior to OIF, they had only emplaced it in a pipeline mode up to a distance of 17 miles. The eventual operation of almost 90 miles total of HRS in Iraq represented the successful implementation of an unproven concept.[7]

To employ its retail system to achieve long-distance fuel delivery, the Marine Corps needed to assemble units and assets from around the globe. Due to the lower-pressure pumps organic to the HRS system, the Marines needed to site a pump station every 2.5 to 3.5 miles along the trace. Each of these pump stations was manned by approximately 15 Marines who ran the pumps, provided local security, and maintained the trace between stations.

The trace from the breach point to Logistics Support Area (LSA) VIPER, located on the Jalibah airfield, was determined with the use of Defense Mapping Agency topographical products that included an elevation profile. Luckily, the trace was relatively flat and the surfaces mostly easy to dig in. This latter point was important in that the method for laying the HRS was to place it into a v-shaped trench dug into the ground. The trench was prepared by a road grader with its blade on an angle; a few locations with more difficult terrain required the use of a ripper to loosen the soil. The trench served two purposes; first, it kept the hoseline relatively straight, which is important because hoses have a tendency to "snake" or move laterally when pressurized, and second, it protected the hoseline from damage due to vehicles driving over it. In spite of this precaution, there were still instances of tanks or AMTRACs driving over and damaging the hose.

The actual laying went quickly. The spools of hoseline conduit were loaded directly onto the backs of 5- and 7-ton trucks. The trucks straddled the v-shaped trench, and Marines walking

---

6   Scott B. Tardif, "267th Quartermasters in Somalia," *Quartermaster Professional Bulletin*, Winter 1993.

7   Brigadier General Edward G. Usher III, Director of Logistics Plans, Policies, and Strategic Mobility, United States Marine Corps, Testimony Before the House Armed Services Committee, March 30, 2004.

Figure 2.2
HRS Laid in V-Shaped Ditch in Southern Iraq

SOURCE: Photo courtesy of Mr. Joseph Irwin,
Tactical Fuel Systems Project Officer, U.S. Marine
Corps Systems Command, Quantico, Virginia.
RAND *TR652-2.2*

behind the vehicles made sure the hose fell into place as it rolled off the spools. Every so often the Marines would also create a bend in the conduit to allow for expansion. Additionally, they placed some slack into the line to make it easier to effect repairs if the line was damaged (which it was by friendly vehicles inadvertently driving over the hose). The truck/Marine teams were able to lay the hoseline at a rate of roughly five miles per hour. The entire trace took only three days to lay but a total of six days to put into full operation, activities that included emergency repairs, pressure testing, and filling the conduit and intermediate bags along the way. Charging the hoseline occurred as it was emplaced, i.e., filling from the breach point up to the first pump station, then charging to the second pump station, etc. The hoseline system of conduits and intermediate storage bags contained approximately 300,000 gallons alone before fuel came out at the other end.[8]

What the Marines eventually emplaced can be characterized as a combination retail and wholesale system. The lateral distance exceeded 60 miles, as shown in Figure 2.3, typically considered a wholesale distance. And at each of the 17 pump stations along the way, there were

---

[8]   Operational details of laying the Hose Reel System conduit are summarized from an email interview with CWO5 (Ret.) Mike Giambruno, conducted March 27, 2007.

Figure 2.3
Trace of USMC Hose Reel System in Iraq

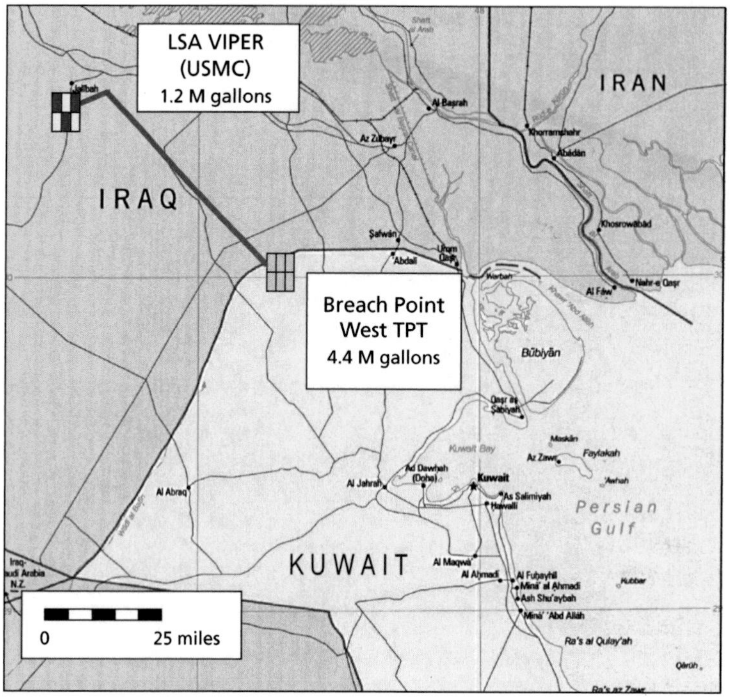

SOURCE: Interviews with USMC personnel.
RAND *TR652-2.3*

two or three 20,000-gallon storage bags, together with an ability to both receive and dispense fuel.

This retail feature (the capability to dispense fuel at each pumping station along the hose-line trace) was also evident in the serial way in which the line was filled, from south to north, pump station by pump station in order. This effective use of retail assets to do both a retail and wholesale mission, though, did have its price. A Bulk Fuel Company was dedicated to its operation for the duration of the mission, and the lines required high maintenance and the constant supervision of an engineer unit that was in support of the fuel mission.[9] In summary, the HRS proved itself a very capable system during the opening phase of OIF. It quickly moved over 8 million gallons of fuel forward, freeing up fuel trucks to concentrate on delivering fuel to the forward edge of battle.[10] But to employ the HRS over a middle to long distance required the massing of equipment and additional personnel dedicated to that mission for its duration.

Army Inland Petroleum Distribution System in OIF

In contrast to the Army pipeline experience in ODS, planning, preparation, and initial construction of the IPDS trace in OIF took place well before the initiation of operations against Iraq. A Quartermaster platoon deployed to Kuwait in 2002 to construct a tactical petroleum terminal at Camp Virginia. In January 2003, the 240th Quartermaster Battalion was deployed

---

[9]   Field Report from Marine Corps Systems Command Liaison Team Central Iraq, April 20 to April 25, 2003.

[10]   Brigadier General Usher testimony.

to Kuwait and began immediate construction of TPTs at the Mina Abdullah refinery, at Camp Udairi, and at Breach Point West near the Iraqi border (Figure 2.4). Also in early 2003, the Army's 62nd Engineer Battalion constructed 51 miles of IPDS pipeline from Camp Virginia through Camp Udairi to Breach Point West. This line was eventually doubled with the construction of a second, parallel line, giving this pipeline trace the capacity to pump up to 1.8 million gallons a day.[11]

After the start of OIF, the IPDS pipeline into Iraq was constructed in three segments. The first of these was from Breach Point West to Jalibah (LSA VIPER) along the same trace as the Marine Corps HRS. It was constructed at an average rate of 2.7 miles per day, with four pump stations emplaced along this segment at roughly 12-mile intervals. The next segment covered 24 miles from VIPER to LSA CEDAR and included two pump stations. Its assembly went much faster, at a rate of 7.2 miles per day. The final segment, 34 miles from LSA CEDAR to the terminus at LSA CEDAR II (Tallil Air Base), incorporated three more pump stations and went in at 2.7 miles per day. This last segment was constructed with IPDS segments

Figure 2.4
Trace of IPDS and Commercial Pipelines in Kuwait and Iraq

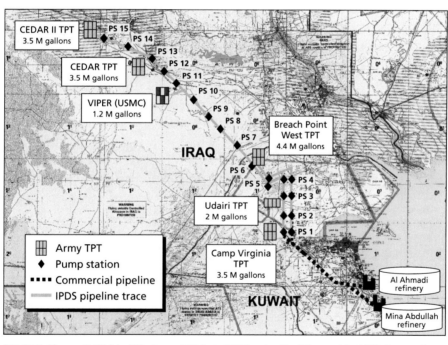

SOURCE: Shawn P. Walsh, *Whatever It Takes: OIF Theater Fuel Support in 2003*, Personal Experience Monograph, Carlisle Barracks, PA: U.S. Army War College, May 1, 2007, p. 2.
RAND TR652-2.4

---

[11] Radian, Inc., 2004, Appendix B, and Colonel Shawn Walsh, *Whatever It Takes: OIF Theater Fuel Support in 2003*, Personal Experience Monograph, Carlisle Barracks, PA: U.S. Army War College, May 1, 2007. A COCOM prewar imperative in OIF was for there to be 11 million gallons in tactical storage prior to crossing into Iraq. Thus it is important to note the role TPTs played in the larger discussion of bulk fuel support. An integral component of the current IPDS unit design, TPTs allowed the tactical fuel truck loading points to be closer to the border, thereby reducing round-trip distances for resupply while still meeting the COCOM's overall on-hand objective and without having to rely on Kuwaiti national storage facilities located a greater distance away near Kuwait City.

recovered from Kuwait. This shift of assets was possible because the IPDS section linking Camps Virginia and Udairi had been replaced by a commercial welded pipe. Once this final section forward was completed, a total of 182 miles of IPDS stretched from Camp Virginia to LSA CEDAR II.[12]

The IPDS system was operated by four Quartermaster companies under the command and control of a Petroleum battalion, but its operation lacked the Military Police (MP) support doctrinally recommended in a nonpermissive environment.[13] The after action report of the Petroleum battalion described significant theft, not only of petroleum products, but also of a considerable quantity of the equipment that carried, pumped, and powered the system.[14]

Providing Drinking Water to the City of Blackstone, Virginia

The historical discussion so far has been confined to uses of pipelines to move petroleum products in combat zones. But the Army's deployable pipeline capability also has the potential for employment in humanitarian relief or homeland security missions to move nonpotable water. An illustration of this capability is support the Army provided to the city of Blackstone, Virginia, in February 2001, when the city water treatment facility discovered a significant leak in its intake. Because the leaking portion of the pipe was encased in concrete, engineers knew that the amount of time required to mend the pipeline would force a disruption of service to customers.

The city of Blackstone negotiated with the Army for the construction of the IPDS system between the reservoir and its water treatment plant (Figure 2.5). Radian, the commercial contractor that maintains the IPDS for the Army at the adjacent Fort Pickett training installation, installed the pipeline along roughly the same trace (about three miles) as the city's permanent pipeline. The original city pipeline was 18 inches in diameter and operated for 8 hours per day. The reduction in pipeline diameter down to the IPDS's 6 inches required that the treatment plant increase its hours of operation to 24 per day for the two days that it took for the city to mend the leak in its own pipeline.[15] But the overall outcome was a very positive one in that there was no disruption in service to city water customers and a savings to local taxpayers.

Summary of Observations from History

Table 2.1 presents a summary of the historical cases cited above. When assembling this chart, it became apparent that one could arrange the cases by distance covered, and in so doing, note that these tend to fall into three distinct groups: (1) short—less than 20 miles; (2) medium—something between 20 and about 50 miles; and (3) long—more than 100 miles. Organizing the instances in this way and then totaling them indicated another pattern, with the shorter

---

[12] Radian, Inc., 2004, Appendix B.

[13] Headquarters, Department of the Army, *Military Police Operations,* Washington, D.C.: Department of the Army, Field Manual 3-19.1, Change 1, January 31, 2002, pp. 4-8 and C-1-2.

[14] 240th Quartermaster Battalion.

[15] "Radian to the Rescue: Pickett Unit to Pump Water While Water Line Is Repaired," *Courier Record,* Blackstone, VA, June 7, 2001., p. A1.

Figure 2.5
IPDS Used to Supply City of Blackstone, Virginia

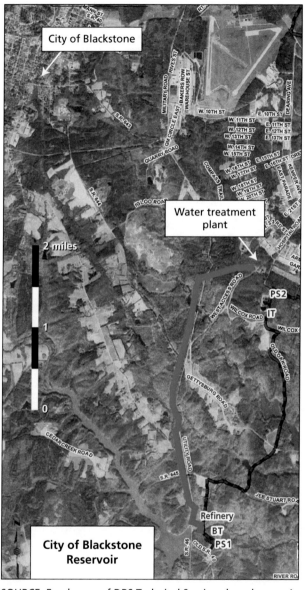

SOURCE: Employees of DRS Technical Services, based on an Army
installation GIS map and information provided during interviews
conducted onsite at Fort Pickett, Virginia, in April 2007.
RAND TR652-2.5

cases being the most frequent at seven or eight (depending if one includes the pipeline constructed but never put into operation in ODS), followed by the medium at three, and one or two long distance (again, this number is dependent on whether one includes the pipeline planned for but only partially constructed in ODS). As the discussion moves next to future pipeline requirements, one will see that this general pattern prevails in planning scenarios: the most numerous pipeline requirements fall into the short-distance category, with some middle-distance requirements, and the rarest being long-distance requirements.

Table 2.1
Summary of Historical Cases from Vietnam to OIF

	Short Distance Up to 25 Miles	Medium Distance ~50 Miles	Long Distance 100 Miles +
Vietnam	DONG NAI–LONG BIN AB • 4 miles: 6″ LWST (3 parallel) POL PIER–CAM RANH BAY AFB • 6 miles: 6″ LWST (2 parallel) SAIGON–TAN SAN NHUT AFB • 6 miles: 6″ LWST (2 parallel) PHAN RANG–PHAN RANG AFB • 10 miles: 6″ LWST (2 parallel) QUI NHON–PHU CAT AFB • 17 miles: 6″ LWST VUNG RO BAY–TUY HOA • 18 miles: 6″ LWST (2 parallel)	QUI NHON–AN KHE • 50 miles: 6″ LWST AN KHE–PLEIKU • 59 miles: 6″ LWST	
Desert Shield	RAS TANURA–KING FAHD APT • 25 miles: 6″ IPDS (contaminated, not used)		ADDAMMAM-HAFIR AL BATIN • 260 miles: 6″ IPDS (partially complete at time of cease fire)
Somalia	MOGADISHU: PORT–AIRFIELD • 2.5 miles: 6″ IPDS		
Iraqi Freedom		USMC: BP WEST–LSA VIPER • 54 miles: 6″ HRS	UDARI–TALLIL • 160 miles: 6″ IPDS
Total	7 (8)	3	1 (2)

NOTES: HRS = Hose Reel System, AB = air base, AFB = Air Force Base, LWST= lightweight steel tubing, IPDS = Inland Petroleum Distribution System.

Potential Future Pipeline Requirements

In order to craft a robust view of potential future pipeline requirements, we reviewed several data sources, many of them classified. The unclassified list shown in Table 2.2 contains enough detail, though, to accomplish the goal of illustrating the breadth and characteristics of planning scenarios included.

The sources of the scenarios were the most recently completed Total Army Analysis (TAA 08–13), COCOM operation plans, and Office of the Secretary of Defense homeland security and steady-state security posture scenario lists. From these sources, there emerged a list of eight instances of either explicit or potential requirements for petroleum pipeline. Each is discussed below, in no particular order except for being roughly grouped into two categories: those scenarios that are more time sensitive from an emplacement standpoint come first, followed by those that are less time sensitive.[16]

---

[16] The focus on time sensitivity as a key measure may be seen as less compelling than other potential measures, such as "gallons delivered per unit time." In the context of this research, though, the alternative newer system the Army was seriously considering, RIFTS, was promoted primarily due to its promised speed of emplacement while holding delivery volume

Table 2.2
Summary of Anticipated Future Requirements

Requirement	Length (miles)	Event	Unclassified Description	Time Sensitive?
A	25 (likely requiring multiple lines)	ISB	Permissive but remote location, time sensitive, very high throughput required	Yes
B	50	MCO	Time-sensitive requirement to move POL	Yes
C	50	MCO	Time-sensitive requirement to move POL	Yes
D	10	NEO	Austere environment, potential requirement to support other nations' forces as well	Yes
E	10	HA/HLD	Austere environment, respond to an environmental disaster	Yes
F	160	MCO	Long distance pipeline, not time sensitive	No
G	35 (likely requiring multiple lines)	MCO	Not time sensitive, very high throughput	No
H	10	MCO	Short distance, not time sensitive	No

The first scenario, A, is the need for a pipeline to move fuel from a port to an airfield in a permissive security environment in a remote location. Since the airfield in this scenario is an intermediate staging base (ISB) for strategic airlift operations, there is a large requirement for daily fuel delivery, on the order of a million gallons per day. Additionally, this ISB must be set up with less than a month's notice, making it the first of the time-sensitive cases. The total amount of pipeline estimated for this scenario is roughly 75 miles.

The next two events on the list, B and C, are both explicit pipeline requirements in support of major combat operations (MCOs). Each instance calls for about fifty miles of pipeline to be emplaced. And both are time-sensitive emplacements that must become operational in order to support the tactical maneuver plan.

The fourth event, D, is a noncombatant evacuation operation (NEO) that is expected to take place in a relatively austere environment that lacks robust commercial petroleum infrastructure. The mission of the pipeline in this case is to move fuel from a ship to a civilian airfield several miles inland. Additionally, the expectation is that U.S. logistics forces will provide fuel support to the evacuation aircraft of allied and friendly nations on this airfield.

The final time-sensitive scenario, E, is a humanitarian assistance (HA) mission to move potable water in conjunction with a water purification unit in an austere environment in response to a natural disaster, such as an earthquake. Water must be moved from a coastline up to ten miles inland in a permissive environment. The performance requirements of this particular scenario would be similar to those of a homeland defense (HLD) mission in providing water or fuel to a community after a natural disaster, such as an earthquake or hurricane.

The last three scenarios in the table are relatively less time sensitive. The first, F, is the longest pipeline on the list at 160 miles and is associated with a major combat operation. Sce-

---

essentially constant. This was to be the overriding "selling point" and a parameter frequently cited in various war plans, which at the same time often do not make reference to a required number of gallons of fuel to be delivered.

nario G is also associated with an MCO, but its 105-mile requirement is the sum of the need for three parallel 35-mile pipelines needed to achieve a high throughput rate. The last instance, H, is for a short pipeline to be constructed on order during an MCO.

Looking at this group of future requirements by distance to be covered, the set of scenarios falls out in a pattern analogous to that seen in the historical cases. The most frequent instances are those that call for short-range pipelines, A, D, E, and H, or four of the eight. The next most frequent is mid-range, B, C, and G, or three of the eight. This leaves the long-range case with a single instance, F. Thus the anticipated future of pipelines looks very much like their past true life employments.

CHAPTER THREE

How Well Do Existing and Future Systems Meet Emerging Needs?

This chapter provides an overview of the technical characteristics of candidate pipeline systems: the IPDS, HRS, RIFTS (still in development), the Assault Hoseline, and the Tactical Water Delivery System. The chapter then moves on to an analysis of the ability of these various systems to meet the set of future requirements.

Inland Petroleum Distribution System (IPDS)

The Army began development of the IPDS after the Vietnam War as a replacement for its LWST pipeline system. Several improvements over LWST were achieved with the design of the IPDS. To begin with, the IPDS 6-inch diameter pipe sections are just 19 feet in length, making them easily fit into standard 20-foot ISO shipping containers.[1] By constructing the pipeline out of aluminum, each segment became light enough to be positioned by hand. Finally, IPDS pipeline coupling is done with one simple pin connector hammered into place as opposed to the two nuts and bolts for each LWST connection.

By current doctrine, an Army horizontal Engineer Company installs the IPDS at a planned installation rate of 2–3 miles per day.[2] Once in place, a PPTO company is the Army unit designed to operate up to 90 miles of IPDS pipeline. Along this 90-mile trace there will be between 6 and 11 pump stations, situated from 8.5 to 15 miles apart, depending upon terrain (closer if going uphill, further apart if going downhill).[3] Emplacement of 90 miles of pipeline would take about a month using the planning factor of 2–3 miles per day. When fully installed, the system runs at 740 psi, delivering 600 to 800 gallons per minute or 850,000 to 1,000,000 gallons per day.[4] The IPDS system was intended for wholesale employment, not only the distribution of fuel but also for its storage, having organic capacity of up to 3.78 million gallons of storage in fabric bags.[5]

---

[1] Radian, Inc., 2004, pp. 16, 20.

[2] Radian, Inc., 2004, Appendix A; and John Roberts, "Stretching the Pipeline," *Technology Today*, Vol. 26, No. 1, Spring 2005, p. 7.

[3] Headquarters, Department of the Army, FM 10-416, *Petroleum Pipeline and Terminal Operating Units*, May 12, 1998, p. 4-1.

[4] Radian, Inc., 2004, pp. 4, 15.

[5] Headquarters, Department of the Army, 1998, p. 4-1.

USMC HRS

The Marine Corps' Hose Reel System (HRS) is also a 6-inch diameter system, but instead of 19-foot-long rigid pipes, it comes in half-mile spools of lay flat hoseline (Figure 3.1).[6] One USMC Bulk Fuel Company is doctrinally expected to be responsible for up to six pump stations and four 5-mile sets (20 miles) of hoseline. However, in actual practice one Bulk Fuel Company wound up operating 58 miles of HRS hoseline at the start of OIF.[7] The anticipated pace of installation is 13.5 miles per day.[8] The HRS operates at about 125 psi, moving 320 to 500 gallons per minute, or between 450,000 and 671,000 gallons per day.[9] The Bulk Fuel Company is similar to IPDS in that it is also capable of both distribution and storage. It has organic capacity to store up to 1,120,000 gallons in five 200,000-gallon parallel tank farms and one 120,000-gallon in-line tank farm.[10]

Figure 3.1
U.S. Marine Corps Hose Reel Conduit Being Emplaced in Iraq

SOURCE: Photo courtesy of Mr. Joe Irwin, U.S. Marine Corps Systems Command, Quantico, Virginia.
RAND TR652-3.1

---

[6]   Radian Inc., 2004, p. 19 and Appendix A.

[7]   U.S. Marine Corps, *Organization of the Marine Corps Forces,* MCRP 5-12D, October 13, 1998, p. 5-38; and Michael Giambruno, email discussion with authors on his personal experience with USMC bulk fuel operations in OIF, March 26, 2007 and April 11, 2007.

[8]   Giambruno, 2007.

[9]   Radian, Inc., 2004, p. 19 and Appendix A.

[10]   U.S. Marine Corps Systems Command, Technical Manual 3835-OI/1A, *Marine Corps Tactical Fuel Systems*, Quantico, VA, July 2005, p. 1-5.

Rapidly Installed Fuel Transfer System (RIFTS)

Similar to the Marines' HRS, the RIFTS is made of 6-inch diameter hoseline.[11] Eight 660-foot hoses are wrapped around one-mile reels and arranged into 50-mile segments.[12] As shown in Figure 3.2, a reel holding 4,000 feet of hoseline is mounted on the back of a HEMTT and the hose laid with the use of an Emplacement and Retrieval Device (ERD), which comprises both the reel and an outrigger arm that swings out from the side of the ERD frame. One RIFTS company is to be responsible for two segments or twenty 5-mile sets totaling 100 miles of RIFTS hoseline.[13] These two segments are expected to be installed in four to six days at an anticipated rate of one mile per hour (20 miles per day).[14]

Pump stations will need to be installed every three to ten miles, again depending on the terrain profile.[15] After the system is emplaced, it would be capable of operating between 500 and 550 psi based on tests of the prototype hoseline produced to date. Research is ongoing to improve the RIFTS hoseline to allow it to achieve internal pressure comparable to the IPDS's 740 psi, with the desired outcome for the RIFTS conduit to be capable of pumping 600–800 gallons per minute or 850,000–1,000,000 gallons per day.[16] Intended as a

Figure 3.2
RIFTS Prototype

SOURCE: U.S. Army Combined Arms Support Command, Sustainment Division, Petroleum and Water, *RIFTS Briefing*, undated.
RAND *TR652-3.2*

---

[11] Roberts, 2005, p. 7.

[12] Radian Inc., 2004, p. 25.

[13] CASCOM, Draft File Description of SRC 10417G000, "RIFTS Company Capabilities and Requirements," Fort Lee, VA, September 19, 2006.

[14] Radian, Inc., 2004, p. 3; and Roberts, 2005, pp. 7–8.

[15] This is based on the operating pressure of 740 psi, which has not been attained yet. Roberts, 2005, p. 7.

[16] Even at these mid-range internal pressures, on the order of 400 psi, the hoseline begins to snake or deform, and the upper-range pressure remains a design goal, Radian, Inc., 2004, pp. 4, 31; and Roberts, 2005, p. 8.

supplement to but not a replacement of the IPDS, the RIFTS unit design does not have organic storage capability.[17]

Assault Hoseline

Two more Army petroleum systems capable of moving large quantities of liquid products, though not doctrinally considered wholesale systems, are included in this discussion. The first of these is the Army's Assault Hoseline system. The Assault Hoseline Team consists of 15 soldiers and is typically a part of a Quartermaster Petroleum Support Company. Equipped with four 2.5-mile sets of 4-inch hoseline, it can cover ten miles total, and with organic 350-gallon-per-minute pumps, it can distribute up to 420,000 gallons of bulk petroleum per day. This team is intended primarily to support tactical airfields and may, when applicable, be used to move fuel from railheads to bulk fuel supply points or from collapsible storage tanks to rail cars. The system has no organic storage capability. One stipulation for the use of the Assault Hoseline Team is that its area of employment must be relatively secure for continuous operations, since its austere manning leaves few soldiers available to patrol or guard its components.[18]

Tactical Water Distribution System (TWDS)

The Tactical Water Distribution System (TWDS) is designed for the tactical movement of potable water over a distance of up to ten miles, depending on terrain. Its main elements consist of a fabric water hoseline, 350-gallon-per-minute pumps, and two 20,000-gallon collapsible fabric bags. The TWDS operating team has 20 soldiers assigned and trained to lay or retrieve the hose and set up pumps and temporary storage bags. These teams normally augment either a Quartermaster Water Purification and Distribution Company or an Augmentation Water Support Company to supplement these units' bulk water distribution capability when operating in a General Support role. The planning rate for deployment of the TWDS hoseline from trucks is 3 miles per hour. Water may be stored in the two organic 20,000-gallon collapsible fabric tanks or chlorinated and distributed to users.[19]

Each of the aforementioned systems is compared and contrasted in Table 3.1.

Categories of Future Requirements

Moving now to the analysis of these systems, Figure 3.3 shows the eight future requirements discussed in Chapter Two mapped into four categories. The first requirement, A, is in support of an Intermediate Staging Base. B and C from the time-sensitive major combat operation are combined, since they are very close in basic characteristics and sequential in their source

---

[17] Radian, Inc., 2004, p. 28.

[18] Tommy G. Smithers, "Quartermaster Transformation and the Supply, Petroleum and Water Missions," *Quartermaster Professional Bulletin*, Winter 2001.

[19] Headquarters, Department of the Army, Field Manual 10-52-1, *Water Supply Point Equipment and Operations*, Washington D.C., June 18, 1991, pp. 6-1 to 6-4.

Table 3.1
Comparison of Pipeline Systems Characteristics

	IPDS	Hose Real System (USMC)	RIFTS	Assault Hoseline	TWDS
Operating unit	PPTO Company	USMC Bulk Fuel Company	RIFTS Company	Hoseline Team	Water Distro Team
Number of soldiers/marines	164	188	164	15	20
Doctrinal unit distance	90 miles	20 miles by doctrine (58 miles traversed in OIF)	100 miles (design goal)	10 miles	10 miles
Section size	5-mile sets	5-mile sets	5-mile sets	2.5-mile sets	10-mile set
Conduit type	Rigid aluminum pipe in 19-foot sections	Lay-flat hose wrapped on a spool in half-mile lengths	Lay-flat hose in 660-foot lengths on a reel	Lay-flat hose on a spool	Lay-flat hose on a spool
Diameter	6 inches	6 inches	6 inches	4 inches	6 inches
Operating pressure	740 psi	125 psi	<500–550 psi (in testing so far)	125 psi	150 psi
Pump capacity	800 GPM	600 GPM	800 GPM	350 GPM	600 GPM
Fuel delivery rate	800–1,000K gallons/day	450–671K gallons/day	800–1,000K gallons/day*	180K gallons/day	420K gallons/day
Pump station interval	8.5 to 15 miles	Up to 3.5 miles	3 to 10 miles	Up to 3 miles	Up to 4 miles
Time to emplace	30 days	2 days	4 to 6 days	1 day	1 day
Organic storage capability	3.8 to 21.0M gallons	Up to 1.2M gallons (5 × 200K + 1 x 120K)	None	None	40K gallons

*These figures for RIFTS delivery rate represent a design goal that assumes the system can eventually achieve 740 psi.

SOURCES: Data are drawn from Radian, Inc., 2004; Roberts, 2005, pp. 6–9; CASCOM, Draft File Description of SRC 10417G000, "RIFTS Company Capabilities and Requirements," Fort Lee, VA, September 10, 2006; Department of the Army, FM 10-416, *Petroleum Pipelines and Terminal Operating Units,* Washington, D.C., May 12, 1998; and U.S. Marine Corps Systems Command, Technical Manual 3835-OI/1A, *Marine Corps Tactical Fuel Systems,* Quantico, VA, July 2005.

planning scenarios. The next pair, D and E, noncombatant evacuation operation or humanitarian assistance, instead of being combined are regarded as interchangeable because their short-notice circumstances and demands for fluid transfer are quite similar. The last requirement, F, is a less time-sensitive MCO instance.

This set of scenarios covers a variety of requirement characteristics: time sensitivity, mission duration, distance to be traversed, and fuel throughput. With distance, for example, the range among the scenarios is from 10 to 160 miles. Similarly, along the dimension of mission duration, the interval goes from a few days to weeks. Using this set to evaluate candidate pipeline systems should produce a solution capable of meeting these categories of scenarios or other, similar ones not yet envisioned.

Figure 3.3
Four Scenarios for Analysis

Requirement	Length (miles)	Event	
A	75 (3 × 25)	ISB	1. Intermediate staging base
B	50	MCO	2. Time-sensitive MCO
C	50	MCO	
D	10	NEO	3. NEO *or* Humanitarian Assistance
E	10	HA/HLD	
F	160	MCO	4. Non-time-sensitive MCO
G	105 (3 × 35)	MCO	
H	10	MCO	

RAND *TR652-3.3*

Evaluation of Candidate Systems

Each candidate fuel transfer system was assessed against each of the four scenario types and along four performance dimensions: strategic mobility, investment costs, time to bring into operation, and personnel needed to man the systems. The fuel required to be delivered in each scenario was held constant. One additional capability, truck companies with 7,500-gallon tankers, was added to the analysis. Fuel trucks were included in this comparative capabilities analysis because they represent an alternative to tactical pipelines. Further, the 7,500-gallon tanker company was chosen over a 5,000-gallon tanker truck company to make the truck alternative as competitive as possible to give it a reasonable chance within this evaluation. Additionally, the larger tanker is more akin to a wholesale system than the smaller, tactical HEMTT tanker.

Figure 3.4 shows the mapping of fuel transfer options to the four scenario types, providing an initial screening of each one's basic feasibility to each scenario. There are two scenarios for which trucks are considered an infeasible solution. In the ISB case, the road network would likely be too congested to use trucks; additionally, the anticipated number of truck upload and offload facilities at the ISB would not support delivery of enough fuel each day even if enough trucks could drive back and forth each day between the port and the airfield.[20]

With the NEO, the basic planning assumption is that the country in question is sliding toward chaos with eroding central authority. Under these conditions, which point toward a limited deployment footprint, and due to the time constraints, trucks are deemed an infeasible solution. Similarly, if the actual event is a humanitarian assistance effort, roads in the affected area are probably impassible and speed of employment is of the essence to save lives.

_____

[20] For example, twin IPDS pipelines are capable of delivering 1.8 million gallons of fuel in a 20-hour duty day. It would take 240 loads of 7,500-gallon tanker trucks to equal this quantity of fuel. Assuming trucks could operate on a 24-hour duty day, this would require the delivery of one truckload every 6 minutes, continuously, with no down time.

Figure 3.4
Initial Screening for Feasibility

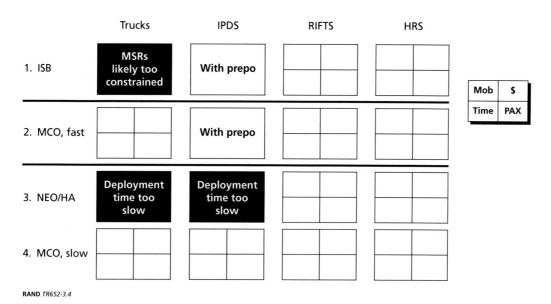

RAND *TR652-3.4*

The IPDS is also considered an infeasible solution for the NEO or HA scenario for comparable reasons. The trace preparation and relatively slow assembly rate of the IPDS simply make it too unresponsive. Also, it is impossible to preposition IPDS sufficiently close to all potential locations to make its use feasible for these NEO and HA scenarios in general. Planners would not know ahead of time where a disaster might strike or which particular government will fall, and likely candidate areas or states do not offer desirable places to store U.S. Army equipment. For two other scenarios, the ISB and MCO fast, IPDS is not considered infeasible but would require strategic prepositioning of assets to be minimally capable of meeting the employment timeline.

Figure 3.5 shows the results of the comparative analysis across the four performance dimensions. As explained above, three of the sixteen quadrants are black because of their infeasibility. For the remainder, looking at the amount of green squares in these thirteen quadrants shows a rough tie between RIFTS and HRS for scenario types 1 through 3. Assuming prepositioning of assets, IPDS does fairly well in scenarios 1 and 2, while trucks do well only in scenario 2. Alternatively, focusing on the red squares in each quadrant, the biggest area of relative weakness for the RIFTS is its cost. The HRS, while relatively cheaper than RIFTS and therefore green in this category, does worse in the other three performance dimensions as the distance to be spanned increases, as is the case with scenario 4.

Summary observations from the analysis are as follows. It appears that trucks and IPDS are roughly comparable. They generally cost less to acquire but are too slow to meet the requirements of one or more scenarios; for those scenarios that they can accommodate, they take longer to employ, especially as delivery distances increase. That said, the IPDS is a proven system and can meet the requirements of three out of the four scenarios if assets are prepositioned in particular regions.

Figure 3.5
Evaluation of Fuel Systems

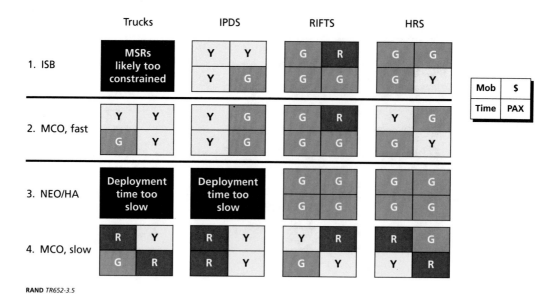

RAND *TR652-3.5*

Moving to the newer technologies, the RIFTS system offers promising performance but remains unproven and has not yet met performance standards in testing. The greatest technical uncertainty is with the hose being able to achieve a high-enough maximum operating pressure to make it possible to achieve throughput comparable to the IPDS. But even if RIFTS does prove able at some point in its development to meet this key performance parameter, it remains by far the most expensive solution to acquire. Much less costly to purchase and proven in combat is the Marine Corps HRS. It does well across all scenarios but will require many more pumping stations and personnel to man them at the longer pipeline distances.[21]

---

[21] The focus on relative acquisition cost is intentional for two reasons. First, the decision where to put the next acquisition dollar was one of the major issues facing the sponsor of this research. Second, the sustainment cost of the various pipeline systems is roughly similar: the conduit elements have a relatively long shelf life, leaving the major cost driver the periodic overhaul and maintenance of the pump stations, pipeline system components that would be needed no matter which conduit system was selected.

CHAPTER FOUR

Policy Recommendations

No Obvious Best Solution

As explained in the previous chapter, the technical assessment does not point to an obvious best solution on what path to take with the future of Army pipelines. In this circumstance, one way to proceed is for the decision maker to apply an appraisal of the external environment in which the decision is being made in conjunction with value judgments that weight the performance dimensions. For example, if cost is deemed to be the most pressing concern, then the preferred path forward may be to go with the system(s) that did the best in the cost category, the IPDS or maybe the HRS. Similarly, if strategic mobility is most important, then the RIFTS (or a program of prepositioning of IPDS assets) would be the most attractive solution.

Table 4.1
Choice in Light of Most Pressing Concern

If highest concern is . . .	. . . then choose
Cost	IPDS (maybe HRS)
Mobility	RIFTS (or prepo IPDS)
Time to employ	RIFTS
Personnel	IPDS + RIFTS
Technical risk	IPDS + trucks

If the decision maker discounts its technical risk and finds speed of employment the most important variable, then RIFTS is the best. Or else if technical risk weighs heavily in the decision, the two proven systems are the IPDS and tanker trucks. Finally, if the number of personnel required to man the systems is the key deciding factor, the IPDS and RIFTS do the best, predominantly when the delivery distance increases and the greater maximum pressure allows for fewer pump stations and hence requires fewer assets to emplace, operate, and guard the pipeline.

No Apparent Rotational Requirement

Temporary pipelines are a very useful capability for a window of requirements. Unlike many support requirements, though, they are not needed for missions that require rotational deployments. It is typically not cost-effective to employ a pipeline unit if the fuel requirement is small or infrequently required. Special cases, such as short-duration noncombatant evacuation or humanitarian assistance missions, are an exception to this rule, as discussed in the previous chapter. But one aspect of such short missions is a second key consideration for the employment of pipelines. It is not desirable to keep them in place too long. As these are above-ground systems, they pose an obstacle to maneuver of both military and indigenous traffic. In areas where military or government control is weak, the pipeline presents an inviting target for either enemy mischief or theft and pilferage. And by their temporary nature in design and materials and due to continued exposure to the elements, they degrade over time, leaking or failing.[1] Thus, if there is a case of an enduring need for pipeline fuel delivery, the preferred solution over time is to have contractors replace the temporary military pipe with a welded and buried pipeline. The transition from temporary to welded pipeline is described in the Army's FM 10-67, page 2-9, and was what the Kuwaiti government eventually did with the section of pipeline connecting the coastal Al Amahdi and Mina Abdullah refineries to Camp Virginia, letting a contract to replace the IPDS trace with a commercial buried pipeline, as noted in Radian, Inc. (2004, Appendix B).

Assess Simultaneous Pipeline Requirement

If there indeed is no rotational requirement, the logistics force structure risk to be covered is the requirement for simultaneous pipeline employment. Returning to the list of likely future scenarios, Figure 4.1 indicates that about 500 miles of pipeline capability could cover all of these requirements at the same time. As this is a military-unique capability, and therefore not readily available from commercial sources, if the simultaneous requirement is the most dangerous future, then the Army ought to have a force with at least this many miles of pipeline and units to operate them. Of course, a rational case can be made for lesser totals or for dividing totals by time frame. For example, if the most time-sensitive scenarios as a group are considered to present the most risk, then their subtotal of about 195 miles could be a lesser subrequirement for rapidly emplaceable pipeline capability. If the solution for the Army is to acquire a hoseline-based system, then a reasonable total buy of this new technology could be on the order of 220 miles, which includes 25 miles for training sets. As a part of this strategy, then, legacy IPDS systems can be retained to cover the remaining 275 miles of less time-sensitive contingency requirements.

Potential Role for TWDS and Assault Hoseline Units

There is, too, the capability of the other two tactical systems described in Chapter Three, the Assault Hoseline and the TWDS, to consider. Each of the associated units is sized and equipped to operate up to ten miles of hoseline. These assets could conceivably meet the requirements of the NEO and HA scenarios. The effect would be to further reduce the potential size of a future tactical pipeline requirement. Taking this thought one step further, and considering the success the Marine Corps had in adapting its equivalent of the Assault Hoseline system—HRS—to a wholesale mission in OIF-1, the Army could consider collapsing PPTO Companies, Assault

---

[1] The Marine Corps System Command planning factors for the hoses in its Hose Reel System are 12 years in storage and 24 months in operation. The more often a hose is deployed or moved in those 24 months will most likely degrade it faster, as points of stress or folds lead to the formation of surface cracks. The aluminum pipe in the IPDS system itself is markedly more durable than a flexible hose. It is the rubber o-rings in the connectors every 19 feet of the IPDS trace, however, that are prone to fail and allow leakage, in some cases immediately after installation.

Figure 4.1
Sizing Total Pipeline Requirement by Simultaneity

Requirement	Length (miles)	Event
A	75 (3 × 25)	ISB
B	50	MCO
C	50	MCO
D	10	NEO
E	10	HA/HLD
F	160	MCO
G	105 (3 × 35)	MCO
H	10	MCO

195 miles (more time sensitive)

275 miles (less time sensitive)

~195 + training set = 220-mile RIFTS buy

~700 miles of IPDS today

470 miles (if all occurred simultaneously)

RAND *TR652-4.1*

Hoseline Teams, and TWDS units into a single unit design with standardized equipment. There are 16 TWDS units in the Army today, 2 Active and 14 Reserve Component, with a total of 320 personnel. There are 16 Assault Hoseline Teams, three Active and 13 Reserve Component, with 240 personnel; adding both unit types yields 560 personnel spaces.

Near-Term Steps

As stated above, the analysis did not point to a clear winner; however, there are two recommended near-term steps the Army could take.

Improved Pump Stations

No matter which path is chosen, though, there are two steps the Army should consider taking today. The first is the investment in an improved pump station. Whether a rigid pipe (IPDS) or flexible hose conduit (RIFTS or HRS) is employed, all require pump stations to operate. The pump stations are themselves systems, consisting generally of a diesel engine, a transmission and clutch, and a pump all mounted on an ISO-compatible platform. The current 800 gpm pumps standard with the IPDS were acquired in the 1980s from a foreign manufacturer and are the most difficult component to maintain. These pumps were a one-time procurement, and spare parts are no longer stocked or produced by the manufacturer. The current estimate of the cost to reset these legacy pumps is $50,000 dollars each. But this reset will also entail the reverse engineering and fabrication of some replacement parts, an aspect that casts some uncertainty on the firmness of the $50,000 reset cost estimate.

Assuming $50,000 per pump, the planned reset of 42 IPDS pumps would total $2.1 million. For roughly the same investment, ten new (RIFTS system) pump stations could be acquired, based on a program estimate of $200,000 per station. The operational advantage of the newer stations is that they create less vibration, which in turn reduces the requirement for engineering support to level the ground at each pump emplacement along the pipeline trace.

Modify PPTO Company

The second near-term step to consider is a modification to the PPTO Company design to provide two improvements. The first is the ability to respond to pipeline requirements that call for less than the full 90-mile design of the PPTO. As discussed earlier, the most frequent employment of pipelines historically has been twenty miles or less; future planning scenarios follow this same pattern. Therefore, the Army should consider making the PPTO Company modular, for example by dividing it into two equally capable platoons. Such reorganization is summarized in Table 4.2.

The modular platoons indicated retain organic storage capability from the original with the proportional number of soldiers allocated against the terminal and bag farm segments of the system. Keeping to the doctrinal 13-mile distance between pumping stations, each platoon, using the same 9-soldier allocation of the original, could operate up to roughly 50 miles of pipeline trace. Thus, it appears possible that a reorganization of existing assets along these lines could yield two platoons able to respond to a simultaneous requirement for two short-range pipelines.

Second, contemporary Army doctrine calls for Military Police companies to patrol the Main Supply Route (MSR) along which IPDS pipeline would typically be emplaced.[2] In the current operating environment, Military Police units have more tasks to perform than units available, making doubtful their availability to protect a pipeline in a nonpermissive environment. This was the case in OIF, where the 240th Quartermaster Battalion operating the IPDS line to Tallil wound up performing the protection mission for its own pipeline. Recognizing

Table 4.2
Making a Modular PPTO Company

PPTO Company with IPDS
1 company ~90 miles

Miles	Sites	Storage	Pipeline	Maint	Misc
0	Terminal	19		23	38
13	PS1		9		
26	PS2		9		
39	PS3		9		
52	PS4		9		
65	PS5		9		
78	PS6		9		
91	Bag farm	30			
		49	54	23	38	164

"Modular" PPTO Company with IPDS
2 platoons ~50 miles each

Miles	Sites	Storage	Pipeline	Maint	Misc
0	Terminal	9		12	19
13	PS1		9		
26	PS2		9		
39	PS3		9		
52	Bag farm	15			
		24	27	12	19	**82**

Miles	Sites	Storage	Pipeline	Maint	Misc
0	Terminal	9		12	19
13	PS1		9		
26	PS2		9		
39	PS3		9		
52	Bag farm	15			
		24	27	12	19	**82**

---

[2]   Headquarters, Department of the Army, Field Manual 3-19.1, Change 1, pp. 4-8 and C-1-2.

this self-protection condition as a realistic potential need, the second option presented here again holds the number of soldiers assigned to the PPTO Company constant but reduces the pipeline operating distance by about half, from 90 miles to 50. As shown in Table 4.2, the soldier positions freed up from the bag farm would then be realigned to pump stations. Again, it appears that a reorganization of an existing unit structure could produce two platoons more capable of operating and protecting a pipeline.

Alternate reorganization options could make up for shortcomings to the doctrinal PPTO design recognized during employment of pipeline units in OIF-1. Such reorganization could be done within the existing personnel counts and would enable units to be self-protecting and allow for more elements (platoons) to cover the most likely future pipeline requirement: delivery distances of 20 miles or less (Table 4.3). As already mentioned, one might even consider consolidating the PPTO, TWDS, and Assault Hoseline teams into one modular pipeline company design. The 560 personnel currently associated with the TWDS and Assault Hoseline units would be more than enough to create three additional modular PPTO Companies.[3]

Table 4.3
Making a Self-Protecting Modular PPTO Company

PPTO Company with RIFTS conduit
1 company ~50 miles

Miles	Sites	Storage	Pipeline	Maint	Misc	
0	Terminal	19		23	38	
10	PS1		13			
20	PS2		14			
30	PS3		14			
40	PS4		13			
50	Bag farm	30				
		49	54	23	38	**164**

"Modular" PPTO Company with RIFTS conduit
2 platoons × ~50 miles each

Miles	Sites	Storage	Pipeline	Maint	Misc	
0	Terminal		10	12	16	
10	PS1		11			
20	PS2		11			
30	PS3		11			
40	PS4		11			
50	Bag farm					
		0	54	12	16	**82**

Miles	Sites	Storage	Pipeline	Maint	Misc	
0	Terminal		10	12	16	
10	PS1		11			
20	PS2		11			
30	PS3		11			
40	PS4		11			
50	Bag farm					
		0	54	12	16	**82**

No organic storage

---

[3] This discussion of alternative unit designs of the current PPTO Company has focused on its capability to operate a pipeline, paying less importance to its petroleum storage and issue capability. Under current PPTO unit doctrine (FM 10-416, Chapter 4), this company can operate two petroleum tank farms or one Tactical Petroleum Terminal (TPT). Thus one potential implication of the suggested platoon-based design could be a future scenario in which multiple platoons are required to operate a large TPT. Another alternative to make the PPTO design more modular could involve the creation of smaller subsets of the existing PPTO platoons—each with integrated modular maintenance and administrative support—while remaining organic to the PPTO company. This examination of alternative force designs is not intended to be exhaustive but to provide stimulus for the more detailed force development work inherent in the process for modifying TO&E documents.

Concluding Observations

In conclusion, temporary pipelines remain an attractive capability to retain in the force structure. Since temporary pipelines do not appear to represent a rotational-based requirement, a reasonable approach to force sizing is to assess which future scenarios that likely call for pipelines might occur simultaneously and determine the aggregate pipeline quantity that would be needed for these scenarios. Of course, no matter how many miles of pipeline units and equipment are programmed for the future force, these capabilities do require additional resources, such as engineer support, to emplace and operate. Investment in new technology, primarily flexible hoseline and improved pumping stations, can lessen emplacement time and engineering support required. Of course, pipelines must be protected, primarily against pilferage or sabotage. Reorganization of the personnel within the existing pipeline unit can make the company more capable of self-protection, at low cost to the Army.

Data Tables

This appendix provides displays of the method used to complete the assessment of each fuel system to meet the requirements of the four scenarios. As already mentioned in Chapter Three, fuel trucks were considered an infeasible solution for two of the scenarios, the ISB and the humanitarian assistance/noncombatant evacuation operation. Likewise, the IPDS system was also deemed infeasible for the HA/NEO scenario. Otherwise, in each scenario the fuel delivery requirement is held constant among the competing systems. Graphs of the relative performance of each of the competing systems are also included.

It is also important to note that this analysis used a modified fielding of the RIFTS technology. Instead of using the costs and equipment of the entire prototype unit design—which included 40 HEMTT vehicles as prime movers for the flexible conduit and automated pump stations, for example—this analysis looked at the less expensive and more modest implementation of the flexible conduit technology inserted into the IPDS unit design. The same approach was taken with the HRS, with the expense of the insertion of HRS conduit into IPDS being the cost of that alternative. Since the IPDS is a legacy system, its cost in this analysis comes from the FY07 budget to reset roughly two companies' worth of IPDS (165 miles).

Finally, Figure A.5 presents the summary of the four previous tables, color-coded with the best across categories green, the worst red, and those in-between yellow. This is the same figure as the one shown in Chapter Three but with the addition of the values in each cell.

Table A.1
Intermediate Staging Base Scenario

Scenario 1: ISB

Length	75	(3 × 25)

Mobility (# of Containers)

IPDS

Equipment	Containers/Units	#	Total Containers
Fuel unit	24	2	48
5 mile set of pipes	13	15	195
Pump station	6	13	78
Pipeline connection assembly	5	1	5
Pipeline support equipment	5	1	5

*Assumes emplacing one pump station every 12 miles.   331

RIFTS

Equipment	Containers/Units	#	Total Containers
Fuel unit	24	3	72
1 mile set of hose and reel	1	75	75
Pump station	6	8*	48
Connection assembly	5	1	5
Support equipment	5	1	5

*Assumes emplacing one pump station every 10 miles.   205

HRS

Equipment	Containers/Units	#	Total Containers
Fuel unit	24	3	72
1 mile set of hose and reel	1	75	75
Pump station	5	27*	135
Connection assembly	5	1	5
Support equipment	5	1	5

*Assumes emplacing one pump station every 3 miles.   292

Trucks	
Infeasible	

Time to Emplace (days)

Trucks	Infeasible	
IPDS	25	(3 miles/day)
RIFTS	4	(20 miles/day)
HRS	8	(10 miles/day)

Trucks	
Infeasible	

Costs

Trucks	
Infeasible	

IPDS	
Reset cost per 90 miles	
	$47,200,000

RIFTS	
FY07 acquisition cost estimate per 5 miles of conduit $4.2M	
75 miles =	$189,000,000

HRS	
FY02 actual cost per 5 miles of conduit $0.281M	
Inflated to FY07 = $0.315M	
75 miles =	$14,175,000

 # Companies

Trucks	Infeasible	
IPDS	1	(1 per 90 miles)
RIFTS	1	(1 per 90 miles)
HRS	2	(1 per 50 miles)

IPDS	
GPM	750
Lines	1
Hours/day	20
Capacity	1,800,000

RIFTS	
GPM	600
Lines	3
Hours/day	20
Capacity	2,160,000

HRS	
GPM	500
Lines	3
Hours/day	20
Capacity	1,800,000

Figure A.1
ISB Scenario:
No Trucks, Zero Days RSOI for Pipelines (Already Prepositioned)

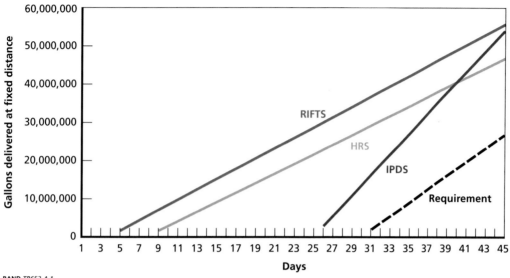

RAND *TR652-A.1*

Table A.2
MCO Fast Scenario

Scenario 2: MCO Fast

Distance	100

Mobility (# of Containers)

IPDS

Equipment	Containers/Units	#	Total Containers
Fuel unit	24	3	72
5 mile set of pipes	13	20	260
Pump station	6	9	54
Pipeline connection assembly	5	1	5
Pipeline support equipment	5	1	5

*Assumes emplacing one pump station every 12 miles.   396

RIFTS

Equipment	Containers/Units	#	Total Containers
Fuel unit	24	3	72
1 mile set of hose and reel	1	100	100
Pump station	6	10*	60
Connection assembly	5	1	5
Support equipment	5	1	5

*Assumes emplacing one pump station every 10 miles.   242

HRS

Equipment	Containers/Units	#	Total Containers
Fuel unit	24	3	72
1 mile set of hose and reel	1	100	100
Pump station	5	34*	170
Connection assembly	5	1	5
Support equipment	5	1	5

*Assumes emplacing one pump station every 3 miles.   352

Trucks

Equipment	Containers/Units	#	Total Containers
Containers/company	205	2.0	410

410

Costs

Trucks

Truck acquisition cost for a 60-truck co.	$15,622,860
Cost for 3 cos.	$31,245,720

IPDS

Reset cost per 90 miles	
	$23,600,000

RIFTS

FY07 acquisition cost estimate per 5 miles of conduit $4.2M	
100 miles =	$84,000,000

HRS (2 parallel lines)

FY02 actual cost per 5 miles of conduit $0.281M	
Inflated to FY07 = $0.315M	
100 miles (×2) =	$12,600,000

 # Companies

Trucks	2	
IPDS	1	(1 per 90 miles)
RIFTS	1	(1 per 90 miles)
HRS	2	(1 per 50 miles)

Time to Emplace (days)

Trucks	9	(mostly RSO&I)
IPDS	33	(3 miles/day)
RIFTS	5	(20 miles/day)
HRS	10	(10 miles/day)

DELIVERY

Trucks

Trucks	80
Capacity	7,500
Speed	20
Distance (round trip)	100
Time/trip	5
Trips/day	2
OR	75%
Capacity	900,000

IPDS

GPM	750
Lines	1
Hours/day	20
Capacity	900,000

RIFTS

GPM	600
Lines	1
Hours/day	20
Capacity	720,000

HRS

GPM	350
Lines	2
Hours/day	20
Capacity	840,000

Figure A.2
MCO Fast Scenario:
No Trucks, Zero Days RSOI for Pipelines (Already Prepositioned)

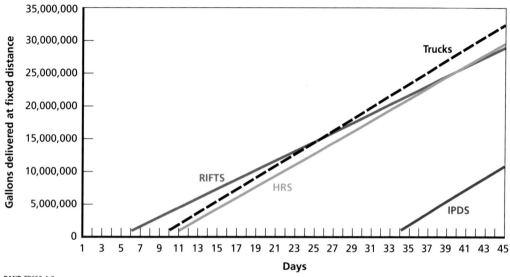

Table A.3
NEO/HA Scenario

Scenario 3: NEO/HA

Distance	10

Mobility (# of Containers)

IPDS

Infeasible	

RIFTS

Equipment	Containers/Units	#	Total Containers
Fuel unit	24	3	72
1 mile set of hose and reel	1	10	10
Pump station	6	1*	6
Connection assembly	5	1	5
Support equipment	5	1	5

*Assumes emplacing one pump station every 10 miles.   98

HRS

Equipment	Containers/Units	#	Total Containers
Fuel unit	24	3	72
1 mile set of hose and reel	1	10	10
Pump station	5	3*	15
Connection assembly	5	1	5
Support equipment	5	1	5

*Assumes emplacing one pump station every 3 miles.   107

Trucks

Infeasible	

Time to Emplace (days)

Trucks	Infeasible	
IPDS	Infeasible	
RIFTS	1	(20 miles/day)
HRS	2	(10 miles/day)

Trucks	
Infeasible	

IPDS	
Infeasible	

RIFTS	
GPM	600
Lines	1
Hours/day	20
Capacity	720,000

HRS	
GPM	350
Lines	1
Hours/day	20
Capacity	420,000

Costs

Trucks

Infeasible	

IPDS

Infeasible	

RIFTS

FY07 acquisition cost estimate per 5 miles of conduit $4.2M	
10 miles =	$8.4

HRS

FY02 actual cost per 5 miles of conduit $0.281M	
Inflated to FY07 = $0.315M	
10 miles =	$0.63

 # Companies

Trucks	Infeasible	
IPDS	Infeasible	
RIFTS	1	(1 per 90 miles)
HRS	1	(1 per 50 miles)

Figure A.3
NEO/HA Scenario:
No Trucks, Zero Days RSOI for Pipelines (Already Prepositioned)

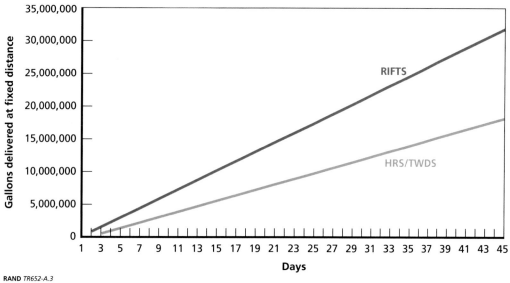

Table A.4
MCO Slow Scenario

Scenario 4: MCO Slow

Distance	100

Mobility (# of Containers)

Costs

IPDS

Equipment	Containers/Units	#	Total Containers
Fuel unit	24	3	72
5 mile set of pipes	13	32	416
Pump station	6	14	84
Pipeline connection assembly	5	1	5
Pipeline support equipment	5	1	5

*Assumes emplacing one pump station every 12 miles.   582

Trucks

Truck acquisition cost for a 60-truck co.	$15,622,860
Cost for 3 cos.	$46,868,580

RIFTS

Equipment	Containers/Units	#	Total Containers
Fuel unit	24	3	72
1 mile set of hose and reel	1	160	160
Pump station	6	16*	96
Connection assembly	5	1	5
Support equipment	5	1	5

*Assumes emplacing one pump station every 10 miles.   338

IPDS

Reset cost per 90 miles	
	$47,200,000

RIFTS

FY07 acquisition cost estimate per 5 miles of conduit $4.2M	
160 miles =	$134,400,000

HRS

Equipment	Containers/Units	#	Total Containers
Fuel unit	24	3	72
1 mile set of hose and reel	1	160	160
Pump station	5	54*	270
Connection assembly	5	1	5
Support equipment	5	1	5

*Assumes emplacing one pump station every 3 miles.   512

HRS (2 parallel lines)

FY02 actual cost per 5 miles of conduit $0.281M	
Inflated to FY07 = $0.315M	
160 miles =	$20,160,000

Trucks

Equipment	Containers/Units	#	Total Containers
Containers/company	205	3	615

615

 # Companies

Trucks	3	
IPDS	2	(1 per 90 miles)
RIFTS	2	(1 per 90 miles)
HRS	4	(1 per 50 miles)

Time to Emplace (days)

Trucks	9	(mostly RSO&I)
IPDS	53	(3 miles/day)
RIFTS	8	(20 miles/day)
HRS	16	(10 miles/day)

DELIVERY

Trucks

Trucks	180
Capacity	7,500
Speed	20
Distance (round trip)	320
Time/trip	16
Trips/day	1
OR	75%
Capacity	1,012,500

IPDS

GPM	750
Lines	1
Hours/day	20
Capacity	900,000

RIFTS

GPM	600
Lines	1
Hours/day	20
Capacity	720,000

HRS

GPM	350
Lines	2
Hours/day	20
Capacity	840,000

Figure A.4
MCO Slow Scenario:
Nine Days RSOI for Trucks, Zero Days RSOI for Pipelines

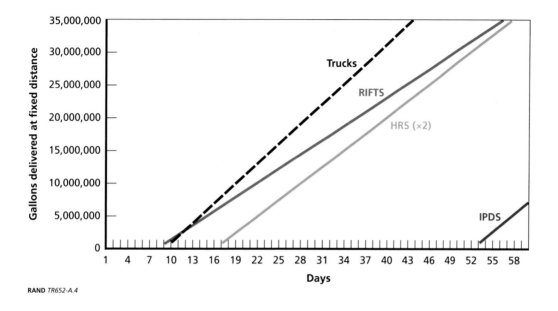

RAND *TR652-A.4*

Figure A.5
Summary Table of Comparative Performance

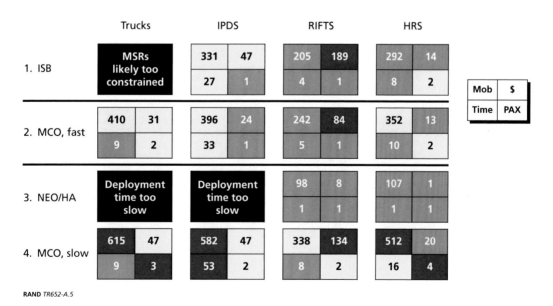

RAND *TR652-A.5*

Bibliography

"Gas Shortage Eases in Phoenix as Pipeline Resumes Pumping," *New York Times,* August 25, 2003. As of February 2009:
http://query.nytimes.com/gst/fullpage.html?res=9C02E7DF1239F936A1575BC0A9659C8B63

Giambruno, Michael, email discussions with the authors, March 26–April 11, 2007.

Headquarters, Department of the Army, Field Manual 10-67, *Petroleum Supply in Theaters of Operations,* Washington, D.C.: Department of the Army, February 18, 1983.

Headquarters, Department of the Army, *Water Supply Point Equipment and Operations,* Washington, D.C.: Department of the Army, June 18, 1991.

Headquarters, Department of the Army, Field Manual 5-482, *Military Petroleum Pipeline Systems,* Washington, D.C.: Department of the Army, August 26, 1994.

Headquarters, Department of the Army, Field Manual 10-416, *Petroleum Pipeline and Terminal Operating Units,* Washington, D.C.: Department of the Army, May 12, 1998.

Headquarters, Department of the Army, Field Manual 3-19.1, Change 1, *Military Police Operations,* Washington, D.C.: Department of the Army, January 31, 2002.

Heiser, Joseph M., Jr., *Vietnam Studies: Logistic Support,* Washington, D.C.: U.S. Government Printing Office, 1991.

"Hoseline System Ensures Steady Fuel Supplies in Uzbekistan," *Army Logistician,* March-April 2003.

Joint Chiefs of Staff, Joint Publication 4-03, *Joint Bulk Petroleum and Water Doctrine,* Washington, D.C.: Joint Chiefs of Staff, May 23, 2003.

"Marines in Iraq Refueled by Logistics Specialists," *Bulk Transporter,* November 1, 2003.

Mattox, Keith E., "The Army's Inland Petroleum Distribution System," *Quartermaster Professional Bulletin,* Spring 1998.

Moon, Steven L., "267th Quartermaster Company at a Joint Forces Exercise in Honduras," *Quartermaster Professional Bulletin,* Summer 2004.

Parsons, Gary L., "Operation Iraqi Freedom Bulk Petroleum Distribution—'Proud To Serve' Style," *Quartermaster Professional Bulletin,* Autumn 2003.

"Radian to the Rescue: Pickett Unit to Pump Water While Water Line Is Repaired," *Courier Record,* Blackstone, VA, June 7, 2001.

Roberts, John, "Stretching the Pipeline," *Technology Today,* Vol. 26, No. 1, Spring 2005.

Smith, Merwin H., "Petroleum Supply in Korea," *Quartermaster Review,* November–December 1951.

Smithers, Tommy G., "Quartermaster Transformation and the Supply, Petroleum and Water Missions," *Quartermaster Professional Bulletin,* Winter 2001.

"Stretching the Pipeline: A New Transfer System Keeps Fast-Moving Armies Supplied with Fuel and Water," *Technology Today,* Spring 2005.

Tardiff, Scott B., "267th Quartermasters in Somalia," *Quartermaster Professional Bulletin,* Winter 1993.

Thomas, Joseph T., *Petroleum Operations in the Gulf War: An Operation Desert Storm Personal Experience Monograph*, Strategy Research Project, Carlisle Barracks, PA: U.S. Army War College, April 15, 1993.

U.S. Marine Corps, *Organization of the Marine Corps Forces,* MCRP 5-12D, October 13, 1998.

U.S. Marine Corps Systems Command, *Marine Corps Tactical Fuel Systems*, Technical Manual 3835-OI/1A, Quantico, VA, July 2005.

Walsh, Shawn P., *Whatever It Takes: OIF Theater Fuel Support in 2003*, Personal Experience Monograph, Carlisle Barracks, PA: U.S. Army War College, May 1, 2007.

Weaver, Kimberly A., *The Inland Petroleum Distribution System (IPDS): Can It Fuel the Force*, Carlisle Barracks, PA: U.S. Army War College, April 10, 2001.